Tony Silveira
Ricardo Robaldo
Ana Luísa Valente

Helminth parasites of Paralonchurus brasiliensis (Steindachner, 1875)

Tony Silveira
Ricardo Robaldo
Ana Luísa Valente

Helminth parasites of Paralonchurus brasiliensis (Steindachner, 1875)

Marine parasitology

ScienciaScripts

Cover image: www.ingimage.com

This book is a translation from the original published under ISBN 978-3-330-77108-6.

Publisher:
Sciencia Scripts
is a trademark of
Dodo Books Indian Ocean Ltd. and OmniScriptum S.R.L publishing group

120 High Road, East Finchley, London, N2 9ED, United Kingdom
Str. Armeneasca 28/1, office 1, Chisinau MD-2012, Republic of Moldova, Europe
Managing Directors: Ieva Konstantinova, Victoria Ursu
info@omniscriptum.com

Printed at: see last page
ISBN: 978-620-8-63558-9

SUMMARY

Chapter 1 2
Chapter 2 22

Chapter 1

Nematode parasites of the maria-luisa, *Paralonchurus brasilensis* (Steindachner, 1875) (Perciformes, Sciaenidae), from southern Brazil

Tony Silveira[1]

Mariana H. Remiao[1]

Pedro Quevedo[1]

Ricardo B. Robaldo[1]

Ana Luisa S. Valente[1]

[1] Graduate Program in Parasitology. Institute of Biology. Federal University of Pelotas. RS, Brazil.

[2] Postgraduate Program in Biotechnology. Center for Technological Development. Federal University of Pelotas. RS, Brazil.

Summary

Sciaenidae fish are the most important components of the ichthyodemersal community in Brazilian coastal waters. The maria-luisa, *Paralonchurus brasiliensis* (Steindachner, 1875), is part of this group of fish. There are two large populations of this species in Brazil. The helminth fauna of the southern population is almost completely unknown. Considering that helminths have been used as markers of dispersal routes, identification of ecological stocks and feeding habits of their hosts, the aim of this study was to characterize parasitism by nematodes in *P. brasiliensis*, contributing to knowledge of the species and the southern population of South America. A total of 43 specimens were necropsied. The fish were caught in a seine net approximately 8km off the coast of the municipality of Rio Grande, Rio Grande do Sul, Brazil, and kept frozen until the time of the parasitological analysis. The parasites collected were fixed in AFA, stained with hematoxylin and clarified in Amman's lactophenol. Nematode parasitism occurred in 100% of the fish analyzed. Among the parasites were larvae *of Hysterothylacium* sp. (P=95.35%; IMI=10.80), the first record for the species; adults and larvae of *Procamallanus (Spirocamallanus) pereirai* (P=32.56%; IMI=1.93) and adults of *Dichelyne (Dichelyne) spinicaudatus* (P=13.95%; IMI=1.5), the first record for the species and for Brazilian waters. *Hysterothylacium* sp. proved to be the best population biomarker, since it was found in the larval stage, with a high prevalence and exclusive occurrence in the southern population. *Dichelyne spinicaudatus* also proved to be an exclusive component of the endoparasite community in the southern population of *P. brasiliensis*. However, it is present in adult hosts and has easy access to the environment through the anal opening, which is why it was characterized as a population biomarker that was not as efficient as *Hysterothylacium* sp. This study of the nematode community of *P. brasiliensis* corroborates the existence of two populations for the species in the southwest Atlantic.

Keywords: Anisakidae, markers, marine, parasites, fish

INTRODUCTION

Sciaenidae species are the most important components of the demersal fish community in the coastal waters of south-eastern and southern Brazil (Soares and Vazzoler 2001). The maria-luisa, *Paralonchurus brasiliensis* (Steindachner, 1875), is included in this group of species. It is distributed from Panama to Argentina, inhabiting estuarine and coastal waters at depths of less than 100m. In Brazil, the species feeds mainly on polychaetes, crustaceans and small fish (Menezes and Figueiredo 1980, Paiva Filho and Rossi 1980, Robert et al. 2007). Based on meristic and morphometric characters, Vargas (1976) identified two populations on the south-east coast of Brazil, one north of 29°S and the other to the south. Paiva Filho and Zani-Teixeira (1980) confirmed the distinction and found seasonal and latitudinal variations in the spatial overlap of the two populations between 22°S and 29°S.

Paralonchurus brasiliensis has low economic value in fishing activities and is considered a "discard" by fishermen in Brazil (Paiva Filho and Schmiegelow, 1986).

However, because individuals of this species represent an important part of the demersal ichthyofauna of the continental shelf of the south-eastern and southern regions of Brazil, the species has great ecological value for the Brazilian coastal ecosystem, being important prey in the diet of various pinnipeds, cetaceans and piscivorous birds (Braga 1990, Waessle et al. 2003).

In recent decades there has been a considerable increase in the attention paid to ichthyoparasitology, as well as the pathology of fish and aquatic organisms (Luque 2004, Silva-Souza 2006). Among the most studied host species are those of some economic interest, due to the intensification of aquaculture activities in Brazil and worldwide (Luque 2004). Thus, studies into the parasitology of freshwater fish have gained greater attention, to the detriment of marine ichthyoparasitology. Parasitic species of marine fish can be used as bioindicators to indirectly assess the status of their environment (Palm and Rückert 2009). In addition, parasitic helminths have been used as markers of dispersal routes, identification of ecological stocks and the diets of their hosts (Mackenzie 1987, Moser 1991). Little is known about the helminth fauna of *P. brasiliensis* in the southern population. Only the Trypanorhyncha cestodes parasitizing the species have been studied in southern Brazil (Pereira Jr. and Boeger 2005). The aim of this study was to report and characterize nematode parasitism in *P. brasiliensis*, contributing to knowledge of the biology of the species and its population in southern Brazil.

MATERIAL AND METHODS

Forty-three specimens of *P. brasiliensis* with an average length of 19.86 ± 1.22 cm and an average weight of 78.23 ± 15.59 g were necropsied. The fish were caught by the industrial fishing fleet in the summer, approximately 8 km off the coast of Cassino beach, Rio Grande, Rio Grande do Sul, Brazil (32°30' S, 52°30' W-33°30' S, 53°30' W), and kept frozen at -20 °C until parasitological analysis. The species was identified according to Menezes and Figueiredo (1980). During necropsy, eyes, gills, heart, stomach, pyloric cecum, intestine, liver, spleen, kidneys, gonads, swim vesicle and musculature were removed and analyzed. Each organ was individualized in Petri dishes, washed in a sieve (150 µm) and examined under a stereomicroscope. Washings from the coelomic cavity and nasal cavity were also analyzed.

The parasites collected were counted and processed according to Amato et al. (1991), fixed in AFA (ethanol, formalin, acetic acid), stained with hematoxylin, clarified in Amman's lactophenol and mounted on permanent Enthelan slides®. The identification of nematode species followed Anderson et al. (2009) and specific articles.

The parasitological indices of prevalence (P); mean intensity of infection (IMI); mean abundance (AM) and variation in intensity of infection (VII) were calculated according to Bush et al. (1997). These measurements are given in millimeters (mm). For morphometric variables, the mean value is presented followed by the range of measurements in brackets, and the number of specimens measured, unless previously announced.

RESULTS

Nematode parasitism was found in 100% of the fish analyzed. The Nematoda parasitizing *P. brasiliensis* include larvae *of Hysterothylacium* Ward et Margath, 1917; larvae and adults of *Procamallanus (Spirocamallanus) pereirai* Annereaux, 1946 and adults of *Dichelyne (Dichelyne) spinicaudatus* Petter, 1974.

Nematoda Rudolphi, 1808

Ascaridoidea Railliet et Henry, 1915

Anisakidae Skrjabin et Karokhin, 1945
***Hysterothylacium* Ward et Margath, 1917**
***Hysterothylacium* sp. (Figure 1)**

Only third-stage larvae *of Hysterothylacium* sp. were found in a total of 443 individuals, with a prevalence of 95.35%. The average intensity of infection was 10.80 (±12.58) and the average abundance was 10.30 (±12.49) larvae per host. The intensity of infection ranged from 1 to 74 parasites per host. The coelomic cavity was the most frequent site of infection, where 66.37% of the larvae were present, followed by the intestinal lumen (13.77%), stomach walls (13.54%), pyloric cecums (3.39%), liver (2.26%) and stomach lumen (0.68%).

Morphometry based on 13 specimens. The L3 larvae had a body measuring 7.442 (1.930 - 14.100) in length and a maximum width of 0.214 (0.075 - 0.450). The labia were poorly developed but well delimited. Larval teeth absent. Lateral wings absent. Excretory pore located near the nerve ring, 0.416 (0.180 - 0.570) from the anterior end. Nerve ring 0.317 (0.158 - 0.400) from anterior end. Oesophagus 0.547 (0.230 - 0.850) long by 0.060 (0.020 - 0.093) wide, with maximum width found in the posterior third of the oesophagus. The length of the esophagus corresponded to 7.67% (5.63% - 11.92%) of the total length of the body. Globose or semi-globose ventricle 0.056 (0.030 - 0.088) long by 0.055 (0.028 - 0.088) wide. Ventricular appendage 0.480 (0.236 - 0.950) by 0.068 (0.027 - 0.150). The esophagus/appendix length ratio was 1.13 (0.890 - 2.000). The intestinal caecum measured 0.181 (0.060 - 0.330) long by 0.039 (0.008 - 0.075) wide. The ventricular appendix was 2.8 (1.6 - 3.9) times larger than the intestinal caecum. The conical tail was 0.125 (0.091 - .210) long and had a mucron at the apex.

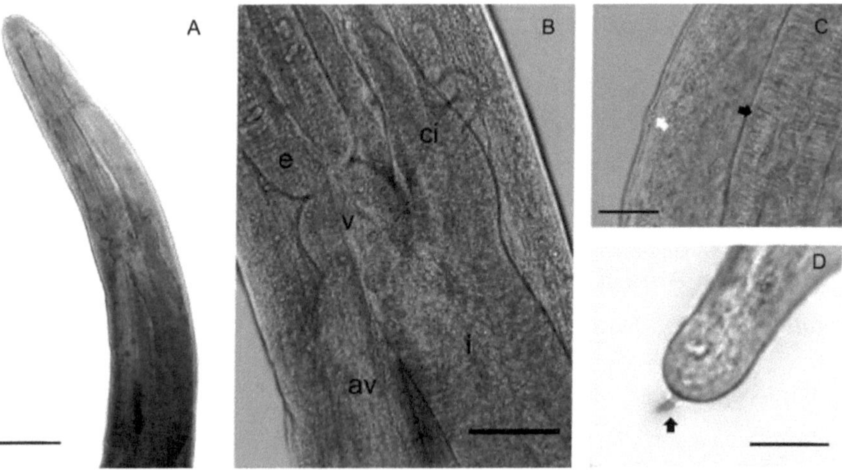

Figure 1. *Hysterothylacium* sp. parasite of *Paralonchurus brasiliensis* (Steindachner, 1875). A: anterior end, bar = 0.2 mm. B: detail of the esophagus/intestine transition, ventricular appendix (av), intestinal cecum (ci); esophagus (e); intestine (i); ventricle (v), bar = 0.07 mm. C: excretory pore (white arrow); esophagus (black arrow), bar = 0.05 mm. D: mucron (black arrow), bar = 0.01 mm.

Camallanoidea Railliet et Henry, 1915

Camallanidae Railliet et Henry, 1915

***Procamallanus* Baylis, 1923**

***Procamallanus (Spirocamallanus) pereirai* Annereaux, 1946 (Figure 2)**

Adults and third and fourth stage larvae, L3 and L4, of *Procamallanus (Spirocamallanus) pereirai* were found in a total of 27 specimens. Prevalence was 32.56%, mean intensity of infection and mean abundance were 1.93 (±1.82) and 0.63 (±1.36) nematodes per host, respectively. The intensity of infection ranged from 1 to 8 parasites per host. The preferred site of infection was the intestine, which was home to 92.59% of all the nematodes recovered. The other 7.41% were represented only by L3 found inside the pyloric cecums.

The L3 of *P. pereirai* had a body measuring 3.291 (2.375 - 4.237) long by 0.057 (0.035 - 0.08) wide, n: 4. Buccal capsule measuring 0.033 (0.028 - 0.036) long by 0.029 (0.024 - 0.035) wide, n: 4, subdivided into an anterior part with 18 to 20 sinuous ridges arranged obliquely, projecting anteriorly to the left and a smooth posterior part, not reached by the ridges. Muscular oesophagus 0.224 (0.193 - 0.255) long by 0.026 (0.020 - 0.034) wide, n: 4. Glandular oesophagus 0.240 (0.170 - 0.310) long by 0.030 (0.018 - 0.038) wide, n: 4. Nerve ring 0.128 (0.113 - 0.138), n: 3, from the anterior end. Tail apex with 3 or 4 spines. Distance between anus and apex of tail 0.085 (0.085 - 0.085), n: 2.

Fourth-stage larvae with a body length of 6.253 (5.725 - 7.275) by 0.108 (0.084 - 0.125) wide, n: 3. Buccal capsule 0.051 (0.049 - 0.053) long by 0.045 (0.043 - 0.048) wide, n: 3, without the subdivision that occurs at L3 and with 12 to 13 sinuous ridges that project anteriorly to the right. Muscular oesophagus 0.371 (0.360 - 0.380) long by 0.039 (0.035 - 0.044) wide, n: 3. Glandular oesophagus 0.400 (0.350 - 0.450) long by 0.039 (0.035 - 0.045) wide, n: 3. Nerve ring 0.201 (0.183 - 0.215), n: 3, from the anterior margin of the body. Tail apex with 2 spines. Distance between anus and apex of tail 0.123 (0.118 - 0.128), n: 2.

Adult males had a body 15.066 (13.230 - 16.930) long by 0.244 (0.205 - 0.280) wide, n: 5. Buccal capsule 0.081 (0.067 - 0.100) long by 0.067 (0.057 - 0.078) wide, n: 5, with 14 to 17 ridges projecting to the right and a conspicuous basal transverse ring. Muscular esophagus 0.442 (0.410 - 0.500) long by 0.065 (0.045 - 0.078) wide, n: 5. Glandular esophagus 0.738 (0.630 - 0.845) long, n: 2, by 0.089 (0.070 - 0.113) wide, n: 3. Nerve ring 0.284 (0.280 - 0.288), n: 2, from the anterior margin of the body. Posterior end of body with caudal wing, 3 precloacal papillae and 6 postcloacal papillae. Unequal spicules, the right one being larger than the left. The largest spicule measured 0.535 (0.460 - 0.600), n: 5, in length. The smaller spicule measured 0.289 (0.220 - 0.358), n: 2. Gubernacle absent. The tail measured 0.212 (0.200 - 0.230) long, with ventral flexion and 2 terminal pointed processes.

Morphometry based on one specimen. Adult female with body measuring 26.500 long by 0.315 wide. Buccal capsule measuring 0.083 long by 0.068 wide, with a basal ring and 14 ridges projecting in the same way as the males. Muscular esophagus measuring 0.500 long by 0.089 wide. Glandular esophagus measuring 0.800 long by 0.110 wide. Nerve ring at 0.268 from the anterior end. Vulva in the posterior half of the body. Single tail with 2 pointed projections homologous to those of the males.

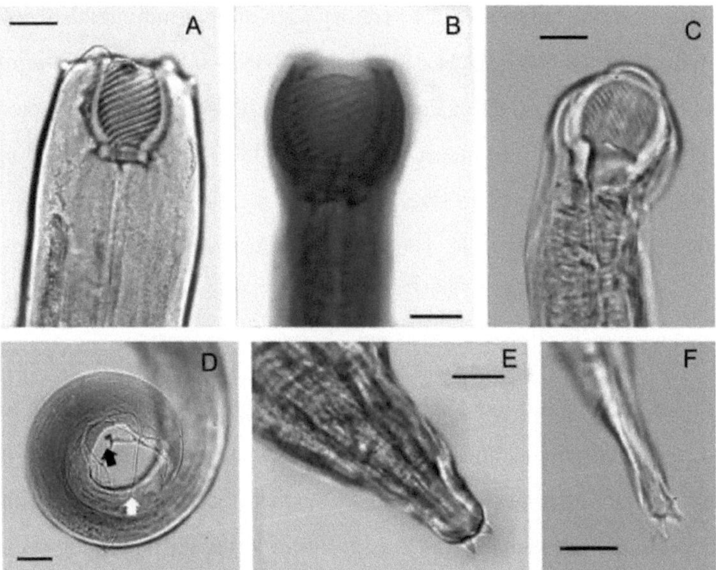

Figure 2: *Procamallanus (Spirocamallanus) pereirai* Annereaux, 1946 parasite of *Paralonchurus brasiliensis* (Steindachner, 1875). A: anterior end of adult male, bar = 0.03 mm. B: anterior end of the fourth-stage larva, bar = 0.015 mm. C: anterior end of third-stage larva, bar = 0.01 mm. D: posterior end of adult male, caudal wing (black arrow); spicule (white arrow), bar= 0.1mm. E: terminal portion of tail of adult female, bar = 0.03 mm. F: terminal portion of the tail of the third-stage larva, bar = 0.01 mm.

Seuratoidea Chabaud, Campana-Rouget et Brygoo, 1959

Cucullanidae Barreto, 1916

***Dichelyne* Jagerskitild, 1902**

***Dichelyne (Dichelyne) spinicaudatus* Petter, 1974 (Figures 3 and 4)**

Adults of *Dichelyne (Dichelyne) spinicaudatus*, totaling 10 individuals, were found. Prevalence was 16.28%, mean intensity and mean abundance of infection were 1.43 (±0.53) and 0.23 (±0.57), respectively. The variation in infection intensity was between 1 and 2 nematodes per host. The site of infection was the intestine.

Morphometric data based on 2 specimens, except gubernaculum measurements (n: 1). Males measured 3.218 (2.660 - 3.775) long by 0.205 (0.180 - 0.230) wide. Nerve ring

0.228 (0.225 - 0.230) from the anterior end. Excretory pore of both individuals 0.500 from the anterior margin of the body. Right deirid 0.440 (0.380 - 0.500) from anterior end and left deirid 0.450 (0.380 - 0.520). Esophagus 0.500 (0.450 - 0.550) long and maximum width in the posterior third measuring 0.084 (0.080 - 0.088). Presence of two intestinal caeca equal or unequal in size, the largest measuring 0.298 (0.245 - 0.350) long by 0.052 (0.041 - 0.063) wide. The right post-deirid is more anterior than the left, 2.001 (1.718 - 2.285) and 2.486 (2.138 - 2.835) from the anterior end, respectively.

Presence of ten pairs of prominent caudal papillae. The first pair ventrally posterior to the left postdeirid, followed almost linearly by the second pair. Third pair situated posterior to the second pair and aligned longitudinally to the preceding pairs, slightly more medial. The fourth pair of caudal papillae is posterior to the third pair and even more medial, being located just anterior to the level of the anterior lip of the cloaca. The fifth pair of papillae is located dorsolaterally and posterior to the fourth pair. The sixth pair of papillae is lateral to the seventh pair when viewed ventrally, and both are posteromedial to the fifth pair. The seventh pair of papillae is the smallest, situated at the base of the anterior lip of the cloaca, these papillae are the closest together. Also on the anterior lip of the cloaca, there is a small odd papilla , which is posterior to the seventh pair of caudal papillae. The eighth pair of papillae is posteromedial compared to the sixth pair and posterolateral compared to the seventh pair. The region around the cloaca has a high density of caudal papillae, comprising pairs IV, V, VI, VII and VIII and the odd papilla on the anterior lip of the cloaca. Posterolaterally to the eighth pair, there is a pair of fasmids that can be mistaken for papillae due to their size and position. Posteromedial to the fasmids is the ninth pair of caudal papillae. Tenth pair of papillae posterolateral to the preceding pair. Spicules sub-equal, right 1.135 (0.910 - 1.360) and left 1.215 (0.950 - 1.480) long. Gubernaculum present, 0.100 long. Tail 0.114 (0.083 - 0.145) long. Precloacal sucker absent.

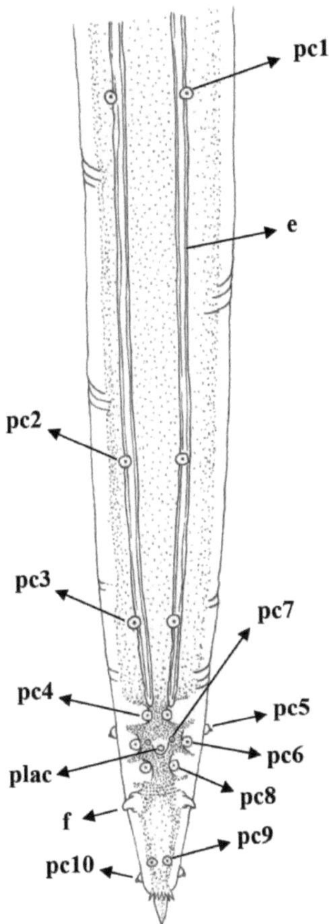

Figura 3. *Dichelyne (Dichelyne) spinicaudatus* Petter, 1974 parasite of *Paralonchurus brasiliensis* (Steindachner, 1875), representation of the posterior end of the male. e: left spicule; f: fasmid; pc1: 1ª left caudal papilla; pc2: 2ª right caudal papilla; pc3: 3rd right caudal papilla; pc4: 4th right caudal papilla; pc5: 5th left caudal papilla; pc6: 6th left caudal papilla; pc7: 7th left caudal papilla; pc8: 8th left caudal papilla; pc9: 9th left caudal papilla; pc10: 10th right caudal papilla.

Females larger than males with body measuring 4.099 (3.612 - 4.687) long by 0.234 (0.170 - 0.285) wide, n: 4. Nerve ring 0.205 (0.135 - 0.275), n: 2, and excretory pore 0.705 (0.480 - 1.000), n: 4, from anterior end. Right deirid 0.519 (0.378 - 0.770), n: 4, distant from anterior end and left deirid at 0.509 (0.385 - 0.720), n: 4. Esophagus 0.585 (0.500 - 0.670) long with widened posterior third of esophagus measuring 0.088 (0.068

- 0.115) wide, n: 4. Presence of two intestinal caeca equal or unequal in size, with the larger one measuring 0.453 (0.395 - 0.510) long by 0.068 (0.050 - 0.085) wide, n: 2. Right post-deirid on the same level as the vulva or anterior to it, 2.532 (1.987 - 3.020), n: 4, from the anterior end. Left post-deirid posterior to the right, distant 3.229 (2.912 - 3.750), n: 3, from the anterior margin. Vulva distant 1.517 (1.230 - 1.963), n: 4, from posterior end. Ovojetor conspicuous, immediately anterior to vulva. Tail 0.124 (0.080 - 0.148), n: 4, long, presence of a pair of fasmids 0.095 (0.093 - 0.098), n: 3, from the apex of the tail. Caudal end with 4 cuticular spines, posterior to the fasmids, followed by an irregular circumferential line of spines, the tail ends in a conical appendage with 4 or 5 spines.

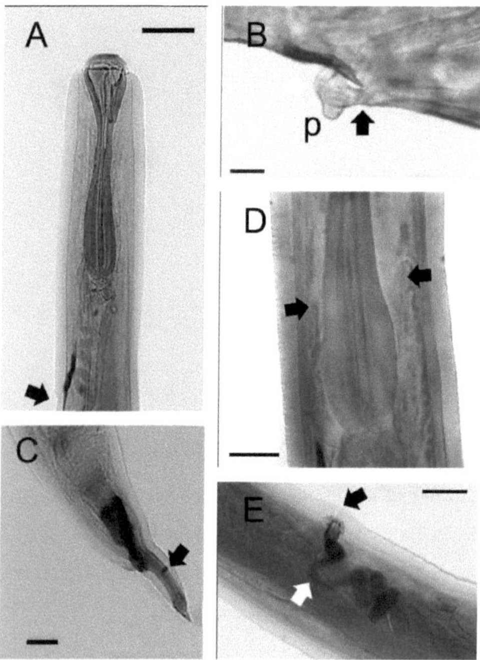

Figura 4. *Dichelyne (Dichelyne) spinicaudatus* Petter, 1974 parasite of *Paralonchurus brasiliensis* (Steindachner, 1875). A: anterior end of female, excretory pore (black arrow), bar = 0.15 mm. B: anterior lip of male cloaca (white arrow) with odd papilla (p); caudal papilla of pair 7 out of focus (black arrow), bar = 0.01 mm; C: posterior end of female, fasmids (black arrow), bar = 0.05 mm. D: intestinal caeca (black arrows), bar = 0.05 mm. E: vulva (black arrow); ovojetor (white

arrow), bar = 0.08 mm.

DISCUSSION

L3 larvae of Anisakidae can be efficiently identified at the generic level (Palmer 2011). However, larval stage individuals of *Hysterothylacium* are easily confused with larvae *of Contracaecum* Railliet and Henry, 1912 (Deardorff and Overstreet 1980). Both genera have an intestinal caecum projecting towards the anterior region of the body and a ventricular appendix projecting posteriorly. Only the position of the excretory pore differentiates the two genera. In *Contracaecum* the excretory pore is located close to the labia, while in *Hysterothylacium* the excretory pore is close to the nerve ring (Deardorff and Overstreet 1980, Deardorff and Overstreet 1981). In their identification key, Deardorff and Overstreet (1981) state that second-stage larvae *of Hysterothylacium* may *already* have a larval tooth. However, the absence of a larval tooth in L3 is possible, as described by Felizardo et al. (2009), Fontenelle et al. (2013) and the present study.

Hysterothylacium species are known to parasitize the digestive tract of fish (Deardorff and Overstreet 1981, 1980). However, these parasites have been reported parasitizing primates (Deardorff and Overstreet 1981), including man (Yagi et al. 1996), so *Hysterothylacium* spp. have zoonotic potential, albeit low. *Hysterothylacium* larvae have already been found parasitizing marine fish off the Brazilian coast. Luque and Poulin (2004) list 20 species of Brazilian marine fish parasitized by larval stages of *Hysterothylacium* spp. *Paralonchurus brasiliensis* does not appear in this list, so this study cites *P. brasiliensis* for the first time as a paratenic host of this anisakid.

Pereira Jr. et al. (2004) made the first record of *Hysterothylacium* larvae on the coast of Rio Grande do Sul, in the same region as this study, in a other host cienid, the corvina, *Micropogonias furnieri* Desmarest. The larvae found in this study were larger than those described by Pereira Jr. et al. (2004). In addition, there was a big difference in the proportion between the length of the ventricular appendix and the intestinal caecum, measuring 2.8 mm in the present study and reaching 5.1 mm in the study by

Pereira Jr. et al. (2004). Another difference is that *Hysterothylacium*, according to Pereira et al. (2004), generally has a larger ventricular appendage than the esophagus, while in the specimens in the present study the opposite is true. These discrepancies could be related to specific differences, since both were only identified at a generic level. The differences could also be related to differences in the host species and the stage of development of the L3, since the intestinal caecum and ventricular appendix can develop at different rates, depending on the age of the L3, as found in *Contracaecum rudolphii* Hartwich, 1964 by Moravec (2009).

Studies along a latitudinal gradient on the Brazilian coast revealed that *Contracaecum* is the only genus of Anisakidae known to date to be part of the parasitic infra-community of *P. brasiliensis* (Luque et al. 2011). Although Luque et al. (2003) did not find *Hysterothylacium* spp. in *P. brasiliensis* off the coast of Rio de Janeiro, Brazil, Cardenas et al. (2012) recorded the occurrence of larvae *of Hysterothylacium* sp. in the stomach and coelomic cavity *of Ctenosciaena gracilicirrhus* Metzelaar (Perciformes: Sciaenidae) with P=10.30% and IMI=1.81, also for the state of Rio de Janeiro, proving the occurrence of *Hysterothylacium* in that region.

The best biomarkers are endoparasites, particularly larval stages, because there is no horizontal transmission and they are almost invariably sequestered in parts of the body without direct access to the external environment, making expulsion impossible (Bush et al., 2001). *Hysterothylacium* sp. showed good characteristics as a biological population marker and its occurrence in *P. brasiliensis*, limited to the southernmost regions of Brazil and the South American continent, reinforces the idea of population division mentioned by Paiva Filho and Zani-Teixeira (1980) and Vargas (1976).

Procamallanus pereirai is a well-known nematode that uses fish as definitive hosts and has been recorded in Brazilian waters. Knoff et al. (2001) reported parasitism by this nematode in *Raja castelnaui* Miranda Ribeiro off the coast of Rio Grande do Sul, Brazil. The studies by Pinto et al. (1984), Santos et al. (1999) and Luque et al. (2003) recorded *P. brasiliensis* as its definitive host, all of which were carried out in the state of Rio de Janeiro. In the present study, in the same way as Santos et al. (1999), larvae and adults of *P. pereirai* occurred simultaneously in *P. brasiliensis,* proving previous statements that this species is an ideal host for *P. pereirai*, since the parasites develop from infective forms to full sexual maturity. Prevalence and abundance were 2.4 and 4.3 times lower, respectively, than those found in the same host species by Santos et al. (1999) on the coast of Rio de Janeiro. Luque et al. (2003) also found higher prevalence and abundance than in the present study, being 1.4 and 3 times higher, respectively. There are no data on nematodes in maria-luisas from the southern population for a proper comparison of the results. However, the lower values of the parasitological indices of the southern population may help to support the idea of two distinct populations of *P. brasiliensis*. Because it is a nematode that reaches sexual maturity in the maria-luisa, *P. pereirai* is a less reliable marker than *Hysterothylacium* sp. (Bush et al., 2001).

Dichelyne spinicaudatus was first reported parasitizing the intestines of *Centropomus undecimalis* Bloch and *Plagioscion auratus* Castelnau in marine waters off French Guiana (Petter 1974). In Argentina, Timi et al. (1997) reported the occurrence of this nematode in *Cynoscion striatus* Cuvier for the first time in the host and in the southwest Atlantic Ocean. The first and only record of *Dichelyne* sp. nematodes in *P. brasiliensis* was made by Pinto et al. (1992) describing the occurrence of *Dichelyne (Cucullanellus) elongatus* (Tornquist, 1931) Petter 1974 off the coast of Rio de Janeiro. *Dichelyne elongatus* and other similar species were taxonomically relocated to *Dichelyne (Cucullanellus) sciaenidicola* Timi, Lanfranchi, Tavares & Luque, 2009 (Timi et al. 2009). In this sense, the present study makes the first record of *D. spinicaudatus* for the Brazilian coast and *P. brasiliensis* is reported by the present study as a new host for this nematode. The distribution of *D. spinicaudatus* restricted to the southern population of *P. brasiliensis* suggests that this nematode has potential as a tool for identifying and characterizing populations of *P. brasiliensis*, as is the case with *Hysterothylacium* sp., but the latter has greater potential for this because it is present in larval stages and in higher prevalence. Even though both reach the adult stage in *P.*

brasiliensis, *D. spinicaudatus* proved to be a better biomarker than *P. pereirai* due to its exclusive occurrence in the southern population (Bush et al. 2001).

Petter (1974) in his description of *D. spinicaudatus*, leaves morphological questions about the males of the species unresolved, as his work does not clarify the exact position of the fasmids and only suggests, but does not state, the presence of an odd papilla on the anterior lip of the cloaca. Timi et al. (1997), using electron microscopy images, confirmed the existence of a small papilla on the anterior lip of the cloaca suggested by Petter (1974). Even with high magnification images, Timi et al. (1997) did not explain the position of the pair of fasmids. This paper provides a detailed description of the position of each caudal papilla, as well as the tail fasmids *of D. spinicaudatus* males.

Both the northern and southern populations have an assembly of nematodes composed of only three species, of which only *Procamallanus* is common in both populations, but with lower levels of infection in the southern population. *Hysterothylacium* sp. turned out to be the best biomarker, since it was found in the larval stage and occurs exclusively in the southern population. *Dichelyne spinicaudatus* also proved to be an exclusive component of the endoparasite community in the southern population of *P. brasiliensis*. However, it has a low prevalence, is present in adult hosts and has easy access to the environment through the anal opening, which is why it was characterized as a biomarker not as efficient as *Hysterothylacium* sp. *Procamallanus pereirai* is not exclusive to the southern population and, even though it is found in larval forms, the species reaches the adult form in *P. brasiliensis*. Thus, the nematode did not prove to be as efficient a population biomarker for the southern lionfish as the other two species studied. However, the comparison between the parasitological indices of this parasite between the southern and northern populations suggests differences between the biology of the two groups. *Paralonchurus brasiliensis* is a new host record for *Hysterothylacium* sp. and *D. spinicaudatus*. The present study of the nematodes of *P. brasiliensis* indicates the presence of two new parasite records for the species and reinforces Vargas' proposal

(1976) and Paiva Filho and Zani-Teixeira (1980) of the existence of two distinct populations for this host in the southwest Atlantic.

REFERENCES

Amato J.F.R., Boeger W.A.E, Amato S.B. 1991: [Laboratory protocols: collection and processing of fish parasites]. Rio de Janeiro: University Press, Federal University of Rio de Janeiro, 81 pp.

Anderson R.C., Chabaud A.G., Willmott S. 2009: Keys to the nematode parasites of vertebrates. London: Cab International. 463 pp.

Braga F.M.S. 1990: [Study of *Paralonchurus brasiliensis* (Teleostei, Sciaenidae) mortality in shrimp-seven beards *(Xiphopenaeus kroyeri)* fishery area]. Bol Inst Pesca. 17: 27-35. (In Portuguese.)

Bush A.O., Fernàndez J.C., Esch G.W., Seed R. 2001: Parasitism: the diversity and ecology of animal parasites. New York: Cambridge University Press, 566 pp.

Bush A.O., Lafferty K.D., Lotz J.M., Shostak A.W. 1997: Parasitology meets ecology on its own terms: Margolis et al. revisited. J Parasitol. 83(4): 575-583.

Càrdenas M.Q., Fernandes B.M.M., Justo M.C.N., Santos A.L., Cohen S.C. 2012: Helminth parasites *of Ctenosciaena gracilicirrhus* (Perciformes: Sciaenidae) from the coast of Angra dos Reis, Rio de Janeiro State, Brazil. Revista Mexicana de Biodiversidad. 83: **31-35**.

Deardorff T.L., Overstreet R.M. 1980: Review of *Hysterothylacium* and *Ihengascaris* (both *previously=Thynnascaris*) (Nematoda: Anisakidae) from the northern Gulf of Mexico. Proc Biol Soc Wash. 93: 1035-1079.

Deardorff T.L., Overstreet R.M. 1981: Larval *Hysterothylacium* (=*Thynnascaris*) (Nematoda: Anisakidae) from fishes and invertebrates in the Gulf of Mexico. Proc Biol Soc Wash. 48: 113-126.

Felizardo N.N., Knoff M., Pinto R.M., Gomes D.C. 2009: Larval anisakid nematodes of the flounder, *Paralichthys isosceles* Jordan, 1890 (Pisces: Teleostei) from Brazil. Neotrop Helminthol. 3: 57-64.

Fontenelle G., Knoff M., Felizardo N.N., Lopes L.M.S., Sao Clemente S.C. 2013: Nematodes of zoonotic importance in *Cynoscion guatucupa* (Pisces) in the state of Rio de Janeiro. Rev Bras Parasit Vet. 22: 281-284.

Knoff M., Sao Clemente S.C., Pinto, R.M., Gomes, D.C. 2001: Nematodes of elasmobranch fishes from the southern coast of Brazil. Mem Inst Oswaldo Cruz. 96: 81-87.

Luque J.L., Alves D.R., Ribeiro R.S. 2003: Community ecology of the metazoan parasites of banded croaker, *Paralonchurus brasiliensis* (Osteichthyes, Sciaenidae), from the coastal zone of the State of Rio de Janeiro, Brazil. Acta Sci, Biol Sci. 25: 273-278.

Luque J.L. 2004: [Biology, epidemiology and control of fishes parasites] Rev Bras Parasitol. 13: 161-165. (In Portuguese.)

Luque J.L., Poulin R. 2004: Use of fish as intermediate hosts by helminth parasites. Acta Parasitol. 49: 353-361.

Luque J.L., Aguiar J.C., Vieira F.M., Gibson D.I., Santos C.P. 2011: Checklist of Nematoda associated with the fishes of Brazil. Zootaxa. 3082: 1-88.

Mackenzie K. 1987: Parasites as indicators of host populations. J Parasitol. 17: 345352.

Menezes, N.A., Figueiredo, J.L. 1980: [Handbook of marine fishes from Southeastern Brazil. IV Teleostei (3)]. Sao Paulo: Museum of Zoology, University of Sao Paulo, 98 pp.

Moravec F. 2009: Experimental studies on the development of *Contracaecum rudolphii* (Nematoda: Anisakidae) in copepod and fish paratenic host. Folia Parasitol. 56(3): 185-193.

Moser M. 1991: Parasites as biological tags. Parasitol Today. 7: 182-185.

Paiva Filho A.M., Rossi L. 1980: [Study on fecundity and spawning *Paralonchurus brasiliensis* (Steindachner, 1875), SP population (Osteichthyes, Sciaenidae)]. Rev Bras Biol. 40(2): 241-247. (In Portuguese.)

Paiva Filho A.M., Schmiegelow J.M.M. 1986: [Study on ictiofauna bycatch fishery of seven-beards-shrimp (*Xiphopenaeus kroyeri*) near the Bay of Santos, SP. 1. quantitative aspects]. Bol Inst Ocean. 34: 78-85. (In Portuguese.)

Paiva Filho A.M., Zani-Teixeira M.L. 1980: [Study of the spatial overlap of populations *Paralonchurus brasiliensis* (Steindachner, 1875) in south-east coast of Brazil between latitudes 22 ° 10'S and 29 ° 21'S (Osteichthyes, Sciaenidae)]. Rev Bras Biol. 40: 143-148. (In Portuguese.)

Palm H.W., Rückert S. 2009: A new approach to visualize ecosystem health by using parasites. Parasitol Res. 105: 593-553.

Palmer S.R. 2011: Oxford Textbook of Zoonoses: Biology, Clinical Practice, and Public Health Control. New York: Oxford University Press, 884 pp.

Pereira Jr. J., Almeida F.M., Morais N.C.M., Vianna R.T. 2004: *Hysterothylacium* sp. larvae (Nematoda, Anisakidae) in *Micropogonias furnieri* (Sciaenidae) from Rio Grande do Sul Coast, Brazil. Rev Atlânt. 26: 55-60.

Pereira Jr J., Boeger W.A. 2005: Larval tapeworms (Platyhelminthes, Cestoda) from sciaenid fishes of the southern coast of Brazil. Zoosystema. 27: 5-25.

Petter A.J. 1974: [Two new species of Cucullanidae fish parasites in Guiana]. Bull Mus Natn Hist Nat., 3er. ser., 225, Zool. 117: 1459-1467. (In French.)

Pinto R.M., Vicente J.J., Noronha D. 1984: First report of *Ascarophis* van Beneden, 1871: *A. brasiliensis* n. sp. (Nematoda, Ascarophidinae) and *Procamallanus (S.) pereirai* Annereaux, 1946 (Nematoda, Procamallaninae) in South America. Mem Inst Oswaldo Cruz. 79: 491-494.

Pinto R.M., Vicente J.J., Noronha, D. 1992: On some family related parasites (Nematoda, Cucullanidae) from the marine fish *Paralonchurus brasiliensis* (Steindachner, 1875) (Pisces, Ostraciidae). Mem Inst Oswaldo Cruz, 87: 207-212.

Robert M.C., Michels-Souza M.A., Chaves P.T. 2007: [Biology of *Paralonchurus brasiliensis* (Steindachner) (Teleostei, Sciaenidae) on the south coast of the state of Paranà, Brazil]. Rev Bras Zool. 24: 191-198. (In Portuguese.)

Santos C.P., Câardenas M.Q., Lent H. 1999: Studies on *Procamallanus (Spirocamallanus) pereirai* Annereaux, 1946 (Nematoda: Camallanidae), with new host records and new morphological data on the larval stages. Mem Inst Oswaldo Cruz. 94: 635-640.

Silva-Souza A.T. 2006: [Health of aquatic organisms in Brazil] Maringà: Abrapoa, 387 pp. (In Portuguese.)

Soares L.S.H., Vazzoler A.E.A.M. 2001: Diel changes and food and feeding activity of sciaenid fishes from the South-Western Atlantic, Brazil. Rev Bras Biol. 61: 197216.

Timi J.T., Lanfranchi A.L., Tavares L.E.R., Luque J.L. 2009: A new species of *Dichelyne* (Nematoda, Cucullanidae) parasitizing sciaenid fishes from off the South American Atlantic coast. Acta Parasitol. 54: 45-52.

Timi J.T., Navone G.T., Sardella N.H. 1997: First report and biological considerations of *Dichelyne (Dichelyne) spinicaudatus* (Nematoda: Cucullanidae) parasite *of Cynoscion striatus* (Pisces: Sciaenidae) from the South West Atlantic Ocean. Helminthologia. 34: 105-111.

Vargas C.P. 1976: [Study on geographical differentiation of *Paralonchurus brasiliensis* (Steindachner, 1875) between latitudes 23 ° 30'S and 33 °]. Msc.

Dissertation , Oceanographic Institute, University of Sao Paulo. (In Portuguese.)

Waessle J.A., Lasta C.A., Favero M. 2003: Otolith morphology and body size relationships for juvenile Sciaenidae in the Rio de la Plata estuary (35-36°S). Sci Mar.

67: 233-240.

Yagi K., Nagasawa K., Ishikura H, Nagagawa A, Sato N, Kikuchi K, Ishikura H. 1996: Female worm *Hysterothylacium aduncum* excreted from human: a case report. Jpn J Parasitology. 45: 12-23.

Chapter 2

Trematodes parasitizing the maria-luisa, *Paralonchurus brasilensis* (Steindachner, 1875) (Perciformes, Sciaenidae), from southern Brazil

Tony Silveira[1]

Mariana H. Remiao[2]

Hugo Amaral[1]

Ricardo B. Robaldo[1]

Ana Luisa S. Valente[1]

[1] Graduate Program in Parasitology. Institute of Biology. Federal University of Pelotas. RS, Brazil.

[2] Postgraduate Program in Biotechnology. Center for Technological Development. Federal University of Pelotas. RS, Brazil.

Summary

The richness of the component communities of many fish on the Brazilian coast has been scarcely studied. An example of this is *Paralonchurus brasiliensis*, which even though it represents an important part of the demersal ichthyofauna of the Brazilian continental shelf and has great ecological value for the Brazilian coastal ecosystem, little attention has been paid to parasitological studies. The aim of this study was to describe the parasitism by digenetics in *P. brasiliensis* and to report new records of species from the parasitic assembly of trematodes in this host. Forty-three specimens of *P. brasiliensis* (19.9 ± 1.2 cm; 78.2 ± 15.6 g) were necropsied. The fish were caught by the industrial fleet, approximately 8 km off the coast of the municipality of Rio Grande, Rio Grande do Sul, Brazil. Each organ was individualized, washed in a sieve (150 μm) and examined under a stereomicroscope. The trematodes collected were fixed in FFA, stained with hematoxylin or Gomori's trichrome, clarified in beech creosote and mounted on permanent slides. The parasitological indices of prevalence (P), mean intensity of infection (IMI) and mean abundance (AM) were calculated. *Diphterostomum brusinae* (P=65.12%; IMI=16.29; AM=10.60)*;* *Opecoeloides catarinensis* (P=16.28%; IMI=1.57; AM=0.26); *O. stenosomae* (P=9.30%; IMI=1.75; AM=0.16); *Pachycreadium gastrocotylum* (P=2.33%; IMI=1; AM=0.023); *Aponurus laguncula* (P=9.30%; IMI=2; AM=0.19); *A. pyriformis* (P=6.98%; IMI=2; AM=0.14) and *Lecithochirium* sp. (P=2.33%; IMI=1; AM=0.023) were part of the digenetic component assembly in *P. brasiliensis*. This is the first record of *D. brusinae*; *O. stenosomae* and *P. gastrocotylum* on *P. brasiliensis*.

Keywords: atlantic, Digenea, marine, parasites, fish

INTRODUCTION

Studies on Trematoda Rudolphi, 1808 in South America began in the 19th century with some European researchers using material collected in Brazil (Kohn et al. 2007). Even with the apparently extensive history, knowledge about tropical aquatic and terrestrial digenetic parasites is probably underestimated (Bush et al. 2001) and it is believed that the marine environment has about twice as many phyla as tropical forests (Suchanek 1994).

The richness of the component communities of many fish species that are very common along the Brazilian coast is still poorly studied. An example of this is *Paralonchurus brasiliensis* (Steindachner, 1875), which even though it represents an important part of the demersal ichthyofauna of the Brazilian continental shelf and has great ecological value for the Brazilian coastal ecosystem, little attention has been paid to parasitological studies.

Only four species make up the component assembly of digenetics known to date for this host (Braga 1990, Kohn et al. 2007). Parasitological studies that have identified trematodes *of P. brasiliensis* in South America are those by Amato (1983a), Ribeiro et al. (2002) and Luque et al. (2003), all based on the northern population of *P.*

brasiliensis proposed by Vargas (1976) and Paiva Filho and Zani-Teixeira (1980). The aim of this study was to describe and characterize digenetic parasitism in *P. brasiliensis* from southern Brazil, as well as to report new records of species that participate in the parasitic assembly of trematodes in this host.

MATERIAL AND METHODS

Forty-three specimens of *P. brasiliensis* with an average length of 19.86 ± 1.22 cm and an average weight of 78.23 ± 15.59 g were necropsied. The fish were caught by the industrial fishing fleet in the summer, approximately 8 km off the coast of Cassino beach, Rio Grande, Rio Grande do Sul, Brazil (32°14'39.9"S and 52°12'15.5"W) and kept frozen at -20 °C until the time of the parasitological analysis. The species was identified according to Figueiredo and Menezes (1980). Each organ was individualized in Petri dishes, washed in a sieve with a 150 μm mesh opening and examined under a stereomicroscope. The organs analyzed were: eyes, gills, heart, stomach, pyloric cecum, intestine, liver, spleen, kidneys, gonads, swim vesicle, sonator muscle and lateral musculature. Washings from the coelomic cavity and the nasal cavity were also analyzed.

The trematodes collected were counted and processed according to Amato et al. (1991), fixed in AFA (ethanol, formalin, acetic acid), stained with Delafield's hematoxylin or Gomori's trichrome, clarified in beech creosote and mounted on permanent Enthelan slides® . The generic identification of the parasites followed Gibson et al. (2002).

The parasitological indices, Prevalence (P); mean intensity of infection (IMI); mean abundance (AM) and variation in intensity of infection (VII) were calculated according to Bush et al. (1997). The measurements mentioned are presented in millimeters (mm), the average value is accompanied by the range of measurements in parentheses, and followed by the number of specimens measured, unless previously announced . For the purposes of morphological description and morphometry, the *hindbody* of the trematodes was considered to begin at the anterior margin of the acetabulum.

RESULTS

Fish infected by trematodes accounted for 69.77% of all hosts analyzed. The endoparasitic digeneans found were *Diphterostomum brusinae* (Stossich, 1888) Stossich, 1903; *Opecoeloides catarinensis* Amato, 1983; *Opecoeloides stenosomae* Amato, 1983; *Pachycreadium gastrocotylum* (Manter, 1940) Manter, 1954; *Aponurus laguncula* Looss, 1907; *Aponurus piryformis* (Linton, 1910) Overstreet, 1973 and *Lecithochirium* sp. The parasitological indices and morphological descriptions of the species are presented below.

Trematoda Rudolphi, 1808

Digenea Carus, 1863

Microphalloidea Ward, 1901

Zoogonidae Odhner, 1902
Zoogoninae Odhner, 1902
***Diphterostomum* Stossich, 1903**
***Diphterostomum brusinae* (Stossich, 1888) Stossich, 1903 (Figure 1)**

Prevalence: 65.12%.

Average intensity of infection: 16.29 (± 21.06) specimens/host.

Average abundance: 10.60 (± 18.63) specimens/host.

Variation in infection intensity: 1 - 95 specimens/host.

Sites of infection: intestinal lumen and pyloric cecum, with 58.55% and 41.45% of digenetics, respectively.

Morphometrics based on 6 specimens. Digenetics with fusiform body measuring 0.888 (0.690 - 1.020) long by 0.193 (0.155 - 0.230) wide. *Forebody* measuring 0.471 (0.320 - 0.600) and *hindbody* 0.393 (0.145 - 0.495) long. It was not possible to see the integumentary spines. Subterminal oral sucker 0.103 (0.075 - 0.113) long and 0.105

(0.090 - 0.120) wide. Globular/subglobular pharynx measuring 0.042 (0.033 - 0.048) long by 0.046 (0.038 - 0.050) wide. Esophagus longer than the caeca. Short caecae, which bifurcate near the level of the genital pore and extend to just before the anterior margin of the acetabulum, with a width of 0.029 (0.023 - 0.035). Acetabulum 0.172 (0.143 - 0.215) long by 0.151 (0.125 - 0.163) wide. Marked presence of large anterior and posterior circumacetabular muscle lips which project 0.046 (0.038 - 0.055) from the perimeter of the acetabulum. Ratio between oral suction cup and acetabulum compositions of 1:1.7 (1:1.30 - 1:2.26). Testicles and ovary surrounded and covered by eggs. Testicles oblique, the anterior one on the left and the posterior one on the right. Anterior testicle overlapping or slightly posterior to the posterior margin of the acetabulum. Anterior testicle 0.085 (0.080 - 0.090) long by 0.068 (0.060 - 0.073) wide. Posterior testicle 0.079 (0.070 - 0.098) long by 0.074 (0.070 - 0.078) wide. Cirrus sac long and flexed to the left, hardly reaching the *hindbody,* measuring 0.163 (0.138 - 0.200) in length and 0.05 (0.046 - 0.053) in maximum width, in the posterior third. Bilobed and subequal seminal vesicle with the largest lobe measuring 0.038 (0.035 - 0.040) long and 0.037 (0.035 - 0.038) wide. Cirrus long and sinuous with a width of 0.023 (0.020 - 0.025), but in its distal portion it takes on a width of 0.05 (0.38 - 0.73), forming a terminal widening. Sinister genital pore. Dorsal pre-testicular ovary, tending towards the right side and overlapping the acetabulum, measuring 0.076 (0.068 - 0.088). Vitellaria made up of two subglobular masses measuring 0.046 (0.030 - 0.065) long by 0.038 (0.030 - 0.053) wide. Uterus filling most of the *hindbody*. Eggs 0.034 (0.030 - 0.038) long by 0.015 (0.013 - 0.018) wide. Terminal excretory pore.

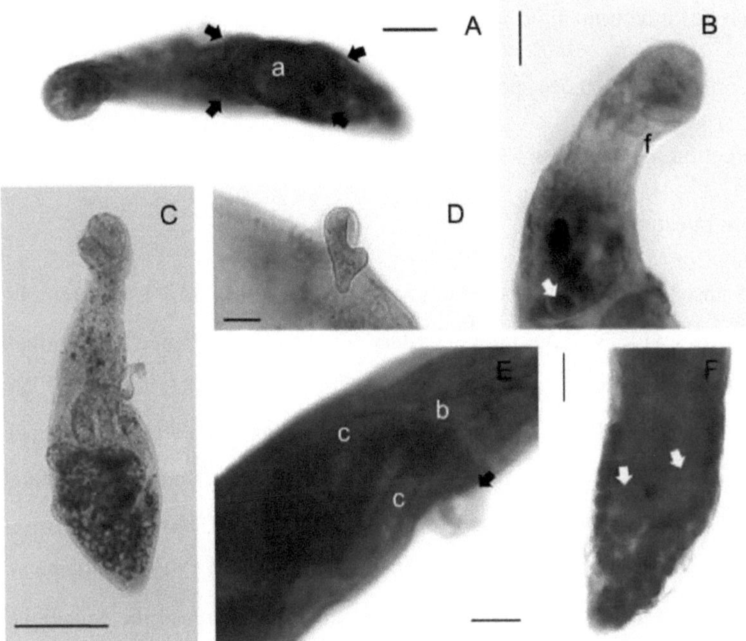

Figure 1. *Diphterostomum brusinae* (Stossich, 1888) Stossich, 1903 parasite of *Paralonchurus brasiliensis* (Steindachner, 1875). A: ventral view of the body, acetabulum (a); muscular lips of the acetabulum (black arrows), bar = 0.1 mm. B: *forebody*, pharynx (f); bipartite seminal vesicle (white arrow), bar = 0.05 mm; C: *in toto* view, unstained specimen, bar = 0.2 mm; D: enlarged terminal portion of cirrus, bar = 0.03 mm; E: ventral view, cecal bifurcation (b); bottom of the cecum (c); cirrus exiting through the genial pore, blurred, (black arrow), bar = 0.04 mm; F: *hindbody*, testicles (white arrows), bar = 0.06 mm.

Allocreadioidea Looss, 1902

Opecoelidae Ozaki, 1925

Opecoelinae Ozaki, 1925

***Opecoeloides* Odhner, 1928**

***Opecoeloides catarinensis* Amato, 1983 (Figure 2)**

Prevalence: 16.28%.

Average intensity of infection: 1.57 (± 0.79) specimen/host,

Average abundance: 0.26 (± 0.66) specimen/host.

Variation in infection intensity: 1 - 3 specimens/host.

Site of infection: intestinal lumen.

Morphometrics based on 3 specimens: Elongated body 1.323 (1.240 - 1.420) long by 0.238 (0.208 - 0.287) wide, without integumentary spines. *Hindbody* much wider than *forebody*. Acetabulum pedunculate, measuring 0.174 (0.165 - 0.184) long by 0.194 (0.191 - 0.198) wide and with 3 papillae on the anterior lip and two wider papillae on the posterior lip. Accessory sucker, sinister, anterior to base of peduncle. Subterminal oral sucker 0.150 (0.116 - 0.183) long by 0.129 (0.125 - 0.132) wide. Short pre-pharynx. Large pharynx, 0.134 (0.125 - 0.142) long by 0.136 (0.120 - 0.159) wide. Esophagus longer than the pharynx. Bifurcated cecums at the level of the acetabular peduncle, extending to the excretory vesicle where they join to form the cloaca. Ratio between widths of oral sucker and acetabulum of 1:1.51 (1:1.45 - 1:1.58). Rounded testicles arranged sequentially in the middle portion of the *hindbody*. Anterior testicle measuring 0.109 (0.100 - 0.115) long by 0.137 (0.122 - 0.156) wide and posterior testicle measuring 0.122 (0.107 - 0.137) by 0.129 (0.111 - 0.151). Narrow tubular seminal vesicle, with the bottom approximately at the level of half the distance between the posterior margin of the peduncle of the acetabulum and the ovary. Ovary also rounded, pre-testicular, measuring 0.079 (0.066 - 0.103) long by 0.067 (0.048 - 0.077) wide. Follicle-shaped vitelline masses distributed circumcecally, extending from the anterior margin of the ovary to the posterior end of the *hindbody*. *Coiled* uterus distributed between the ovary and the peduncle of the acetabulum. Sinister genital pore at the level of the pharynx. Eggs measuring 0.051 (0.047 - 0.054) long by 0.027 (0.025 - 0.028) wide.

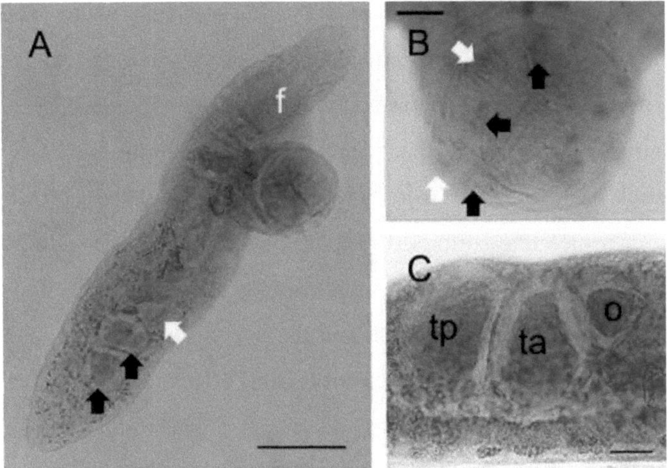

Figure 2. *Opecoeloides catarinensis* Amato, 1983 parasite of *Paralonchurus brasiliensis* (Steindachner, 1875). A: *in toto* view, pharynx (f); ovary (white arrow); testicles (black arrows), bar = 0.2 mm; B: acetabulum, anterior papillae (black arrows) and posterior papillae (white arrows), bar = 0.05 mm; C: ovary (o); anterior testicle (ta); posterior testicle (tp), bar = 0.05 mm.

Opecoeloides stenosomae **Amato, 1983 (Figure 3)**

Prevalence: 9.30%.

Average intensity of infection: 1.75 (± 0.50) specimen/host.

Average abundance: 0.16 (± 0.53) specimen/host.

Variation in infection intensity from 1 to 2 specimens/host.

Site of infection: intestinal lumen.

Morphometrics based on 2 specimens. Narrow, elongated body measuring 1.395 (1.320 - 1.470) long by 0.195 (0.190 - 0.200) wide. *Hindbody* of uniform width. Pedunculate acetabulum, measuring 0.136 (0.107 - 0.164) long by 0.130 (0.102 - 0.157) wide with the presence of 3 papillae on the anterior lip and three papillae on the posterior lip. Accessory sucker between the peduncle of the acetabulum and the genital pore. Subterminal oral sucker 0.018 (0.072 - 0.164) long by 0.104 (0.072 - 0.135) wide.

Short pre-pharynx. Pharynx 0.126 long by 0.128 wide, n:1. Esophagus longer than pharynx. Cecal bifurcation at the level of the peduncle of the acetabulum, joining the excretory vesicle to form the cloaca. Ratio between the width of the suckers of 1:1.29 (1:1.16 - 1:1.42). Rounded testicles arranged sequentially in the posterior half of the *hindbody*. Anterior testicle measuring 0.110 (0.095 - 0.125) long by 0.099 (0.098 - 0.100) wide and posterior testicle measuring 0.101 (0.099 - 0.104) by 0.107 (0.097 - 0.117). Tubular seminal vesicle, with the bottom approximately at the level of half the distance between the peduncle of the acetabulum and the ovary. Ovary round, pre-testicular, measuring 0.078 long in both specimens and 0.061 (0.056 - 0.066) wide. Follicle-shaped vitelline masses with circumcecal distribution, extending from the anterior margin of the ovary to the posterior end of the *hindbody*. Pre-ovarian uterus. Genital pore at the level of the pharynx. Eggs 0.051 (0.048 - 0.053) long by 0.026 (0.023 - 0.028) wide.

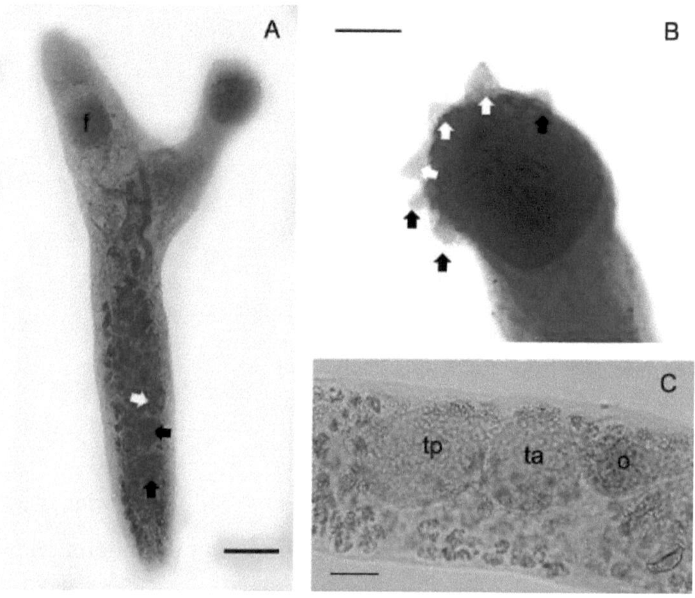

Figure 3. *Opecoeloides stenosomae* Amato, 1983 parasite of *Paralonchurus brasiliensis* (Steindachner, 1875). A: *in toto* view, pharynx (f); ovary (white arrow); testicles (black arrows), bar = 0.15 mm. B: acetabulum, anterior papillae (black arrows); posterior papillae (white arrows), bar =

0.07 mm. C: ovary (o); anterior testis (ta); posterior testis (tp), bar = 0.05 mm.

Plagioporinae Manter, 1947

***Pachycreadium* Manter, 1954**

***Pachycreadium gastrocotylum* (Manter 1940) Manter, 1954 (Figure 4)**

Prevalence: 2.33%.

Average intensity of infection: 1 specimen/host.

Average abundance: 0.023 specimen/host.

Site of infection: pyloric cecums.

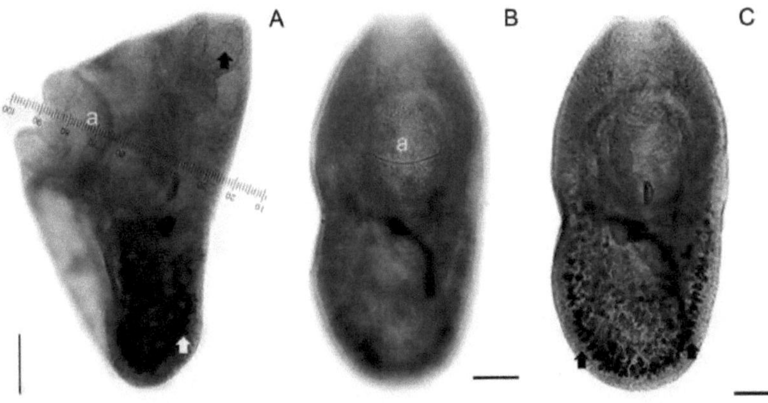

Figure 4: *Pachycreadium gastrocotylum* (Manter 1940) Manter, 1954 parasite of *Paralonchurus brasiliensis* (Steindachner, 1875). A: acetabulum (a); vitelline glands (white arrow); oral sucker (black arrow), bar = 0.2 mm. B: acetabulum (a), bar = 0.15 mm. C: vitelline glands (black arrows) bar = 0.15 mm.

Short, oval-shaped body, 1.23 long by 0.059 wide. Oral sucker located in the dorsal third of the body and acetabulum positioned in the most ventral third of the body. Oral sucker 0.180 long by 0.213 wide. Pre-pharynx very short at 0.015 long. Acetabulum measuring 0.360 long by 0.300 wide. Ratio between width of oral suction cup and acetabulum of 1:1.40. Cirrus sac with a maximum width of 0.073. Long, sinuous cirrus. Submedian genital pore, right-handed, located ventral to the oral sucker. Anterior testicle on left side 0.153 long by 0.188 wide. Posterior testicle on the right, 0.183 long and 0.138 wide. Ovaries poorly developed, 0.108 long by 0.103 wide. Follicle-shaped vitelline masses occupying the posterior half of the *hindbody*, more densely arranged near the margins of this region. Only two eggs of the same size were observed, measuring 0.075 long by 0.061 wide.

Hemiuroidea Looss, 1899

Lecithasteridae Odhner, 1905

Lecithasterinae Odhner, 1905

***Aponurus* Looss, 1907**

***Aponurus laguncula* Looss, 1907 (Figure 5)**

Prevalence: 9.30%.

Average intensity of infection: 2 (± 2.00) specimens/host.

Average abundance: 0.19 (± 0.79) specimen/host.

Variation in infection intensity: 1 - 5 specimens/host.

Site of infection: stomach.

Measurements based on 2 specimens. Narrow body 1.525 (1.400 - 1.650) long by 0.225 (0.220 - 0.230) wide. *Forebody* 0.050 (0.470 - 0.530) and *hindbody* 1.016 (0.920 - 1.112) long. Preoral lobe 0.028 (0.025 - 0.030) long. Oral sucker measuring 0.111 (0.110 - 0.112) long by 0.112 (0.110 - 0.114) wide. Pharynx measuring 0.050 long by 0.060 wide in the specimens measured. Acetabulum in the middle third of the body measuring 0.186 (0.179 - 0.192) long by 0.185 (0.179 - 0.190) wide. Ratio between the widths of the oral sucker and acetabulum of 1:1.65 (1:1.57 - 1:1.73). Seminal vesicle width of 0.057 (0.044 - 0.070). *Pars prostatica* measuring 0.116 (0.110 - 0.121) long by 0.074 (0.072 - 0.075) wide. Testicles aligned longitudinally, in the middle third of the *hindbody*. Anterior testicle 0.113 (0.100 - 0.126) long by 0.108 (0.100 - 0.115) wide. Posterior testicle 0.107 (0.100 - 0.113) long by 0.107 (0.100 - 0.114) wide. Post-testicular ovary, aligned longitudinally to the testicles, 0 to 0.006 from the posterior testicle. Ovary 0.095 (0.089 - 0.100) long by 0.088 (0.085 - 0.090) wide. Vitelline glands posterior to the ovary arranged in roughly spherical lobes with a diameter of 0.051 (0.040 - 0.058). Eggs 0.023 (0.022 - 0.024) long by 0.013 (0.012 - 0.014) wide.

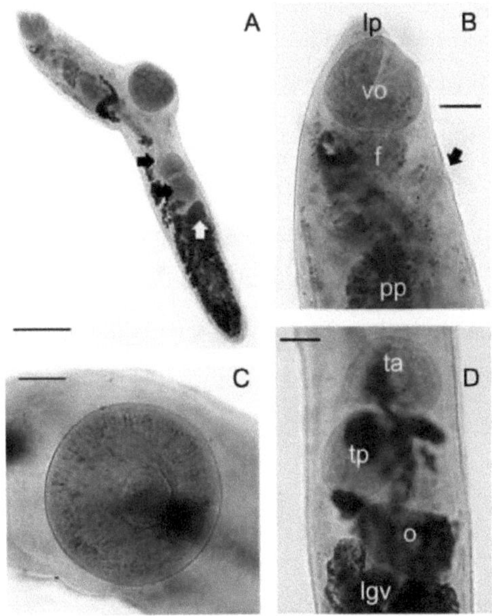

Figura 5. *Aponurus laguncula* Looss, 1907 parasite of *Paralonchurus brasiliensis* (Steindachner, 1875). A: *in toto* view, ovary (white arrow); testicles (black arrows), bar = 0.2 mm. B: anterior end, pharynx (f); pre-oral lobe (lp); *pars prostatica* (pp); genital pore (black arrow), bar = 0.05 mm. C: acetabulum, bar = 0.05 mm. D: ovary (o); vitalinic gland lobe (lgv); anterior testis (ta); posterior testis (tp), bar = 0.05 mm.

Aponurus pyriformis **(Linton, 1910) Overstreet, 1973 (Figure 6)**

Prevalence: 6.98%.

Average intensity of infection: 2 (± 1.73) specimens/host.

Average abundance: 0.14 (± 0.64) specimen/host.

Variation in infection intensity: 1 - 4 specimens/host.

Site of infection: stomach.

Measurements of 1 specimen. Small, fusiform body, 0.720 long by 0.170 wide without integumentary spines. Subterminal oral sucker 0.085 long by 0.088 wide. Presence of a pre-oral lobe. Subglobular pharynx 0.055 long by 0.048 wide. Cecus extending to the posterior end of the *hindbody*. Acetabulum in the middle third of the body 0.155 long by 0.153 wide. Ratio between width of oral sucker and acetabulum 1:1.74. Testicles rounded, symmetrical and immediately post-acetabular. Left testicle slightly more anterior than the right. Left testicle 0.063 long by 0.068 wide. Right testicle measuring 0.065 in both length and width. Seminal vesicle measuring 0.043 long by 0.098 wide. Well-developed *pars prostatica*, anterior to the acetabulum. Genital pore at the level of the pharynx. Post-testicular ovary, in the middle third of the *hindbody*, sinister, rounded and 0.040 long by 0.035 wide. Lobed vitellaria, occupying the middle third of the *hindbody*, lobes join at a point that tends to be positioned slightly to the right of the *hindbody*. Terminal excretory pore and excretory vesicle with arms that join at the level of the posterior half of the oral sucker.

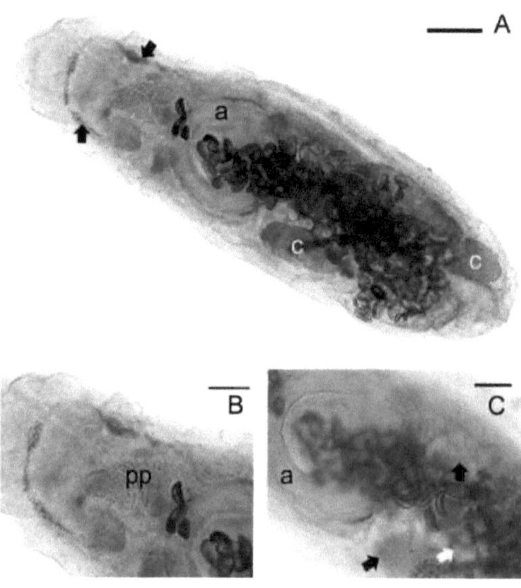

Figura 6. *Aponurus pyriformis* (Linton, 1910) Overstreet, 1973 parasite of *Paralonchurus brasiliensis* (Steindachner, 1875). A: *in toto* view, acetabulum (a); cecum (c); arms of the excretory vesicle (black arrows), bar = 0.1 mm. B: *pars prostatica* (pp), bar = 0.05 mm. C: anterior portion of *hindbody*, acetabulum (a); ovary (white arrow); testicles (black arrows), bar = 0.05 mm.

Hemiuridae Looss, 1899

Lecithochiriinae Lühe, 1901

Lecithochirium Lühe, 1901

Lecithochirium sp. (Figure 7)

Prevalence: 2.33%.

Average intensity of infection: 1 specimen/host.

Average abundance: 0.023 specimen/host.

Variation in intensity of infection: 1 specimen/host. Site of infection: ovary.

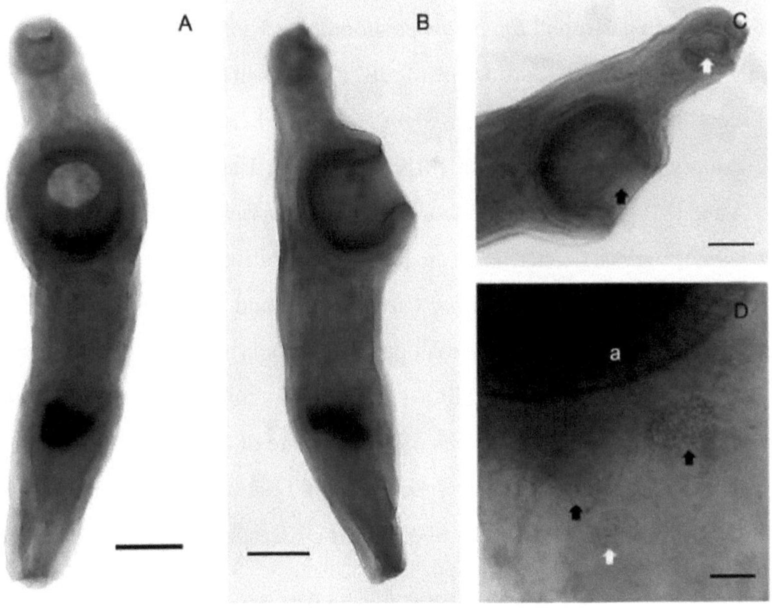

Figure 7. *Lecithochirium* sp. parasite of *Paralonchurus brasiliensis* (Steindachner, 1875). A: *in toto* ventral view, bar = 0.3 mm. B: right lateral view, bar = 0.3 mm. C: acetabulum (black arrow) and oral sucker (white arrow); bar = 0.2 mm; D: posterior edge of acetabulum (a); testicles (black arrows); ovary (white arrow), bar = 0.4 mm.

Measurements based on the only young specimen. Elongated body 3.575 long (with invaginated echoma) by 0.700 wide. Oral sucker 0.300 long by 0.295 wide. Globose pharynx with a diameter of 0.160. Wide and very visible cecum, extending to the posterior end. Acetabulum 0.570 long by 0.610 wide. Ratio between width of oral suction cup and acetabulum 1:2.07. Gonads poorly developed. Symmetrical testicles immediately posterior to the acetabulum. Right testicle 0.083 long by 0.750 wide. Left testicle 0.080 long by 0.105 wide. Median post-testicular ovary 0.450 long by 0.400 wide. Vitellaria and eggs absent.

DISCUSSION

Paralonchurus brasiliensis is recorded in this study as a new definitive host for *Diphterostomum brusinae*. The South Atlantic has already recorded *metacercariae of D. brusinae* in the gastropod *Buccinanops globulosus* Kiener, 1834, in Argentina, with a prevalence of 0.16% (Gilardoni et al. 2011). Also in Argentina, Martorelli et al. (2013) recorded parasitism in *Micropogonias furnieri* Desmarest, 1823 in the Bahia Blanca estuary with P=18.3% and IMI=18.9 and in Bahia de Samborombóm with P=14.7% and IMI=6.7. In Brazil, only *Haemulon steindachneri* Jordan & Gilbert, 1882 and *Orthopristis ruber* Cuvier, 1830 are listed as definitive hosts of *D. brusinae* (Kohn et al. 2007). *Haemulon aurolineatum* Cuvier, 1930 and *Isacia conceptionis* Cuvier, 1930 are recorded as hosts of the trematode in Venezuela and Peru, respectively (Kohn et al. 2007). Bray and Gibson (1986) mention that adult specimens of *D. brusinae* generally occur in the caudal portions of the intestine of bony fish. The results of the present study contradict this, since specimens of *D. brusinae* were recovered from cranial regions of the intestinal tract of *P. brasiliensis*, such as the pyloric caeca.

Opecoeloides catarinensis was described for the first time as a parasite of the intestine of *Micropogonias furnieri*, *Cynoscion leiarchus* Cuvier, 1930 and *P. brasiliensis*, common Sciaenidae fish in southern Brazil (Amato 1983a). The prevalence and average intensity of infection found by Amato (1983a) in *P. brasiliensis* (P=45.45%; IMI=2.4) is higher than that presented in this study and the variation in intensity of infection remained the same, from 1 to 3 parasites per host.

Opecoeloides catarinensis already occurs in the area studied; Pereira Jr. et al. (2000) recorded the trematode in *M. furnieri* on the coast of Rio Grande do Sul. The specimens described in this study fit the measurements described by Amato (1983a), but in general they are smaller than those presented by Pereira Jr. et al. (2000).

Paralonchurus brasiliensis is recorded in this study as a new host of *Opecoeloides stenosomae*. This trematode has already been recorded on the Brazilian coast in *M. furnieri* by Amato (1983a) and Pereira Jr. et al. (2000). Amato (1983a) found a prevalence of 9.52% and an average intensity of infection of 8.5 parasites per host. The parasitism recorded in this study has a higher prevalence and lower average intensity of infection than those recorded by Amato (1983a). In addition, the specimens in this study are smaller than those described by Amato (1983a) and Pereira Jr. et al. (2000).

Paralonchurus brasiliensis is recorded as a new host for *Pachycreadium gcislrocolyliim* in this study. *Pachycreadium gcislrocolylimì* has already been reported in Brazilian waters. Fernandes and Goulart (1992) recorded the species in *M. furnieri* and *Slellifer raslrifer* Jordan, 1889 off the coast of Rio de Janeiro. Pereira Jr. et al. (2000) recorded the species also parasitizing *M. furnieri*, but off the coast of Rio Grande do Sul. The specimen in this study has smaller gonads than those described by Pereira Jr. et al. (2000). This difference in body size could be related to the lack of complete and sexual development of the parasite, since a low quantity of eggs was observed in the specimen. *Pachycreadium gaslrocolylum* described here has a smaller body and gonads than those of *S. raslrifer* studied by Fernandes and Goulart (1992) and all the characteristics described, except the eggs, were smaller than those observed in *M. furnieri* also analyzed by Fernandes and Goulart (1992).

The first record of *Aponurus laguncula* in Brazil was made by Fernandes et al. (1985) on *Chaelodiplerus faber* Broussonet, 1782; *Scomber japonicus* Houttuyn, 1780; *Trachurus lalhami* Nichls, 1920; and *Umbrina coroides* Cuvier, 1830 in the state of Rio de Janeiro. Later, Pereira et al. (2000) reported the species as a parasite of *M. furnieri* in the state of Rio Grande do Sul. Also in Rio de Janeiro, Luque et al. (2003) revealed parasitism in *P. brasiliensis* for the first time, with P=40% and IMI=4.1. The specimens of *P. brasiliensis* collected in this study are equivalent in size to those found by Pereira et al. (2000).

Aponurus pyriformis was recorded by Amato (1983b) off the coast of Santa Catarina, Brazil, parasitizing the stomachs of *Orthopristis ruber* Cuvier, 1880; *Eucinostomus melanopterus* Bleeker, 1863; *Archosargus rhomboidalis* Linnaeus, 1758; and *Isopisthus parvipinnis* Cuvier, 1830. The prevalence found by the author in *I. parvipinnis*, the only cienid infected by *A. pyriformis* in the sample, was 12.5% and the average intensity of infection was 1 parasite per fish. *Haemulon sciurus* Shaw, 1803 was identified as a host for *A. pyriformes* (P = 22.5% and IMI = 1.21) in Rio de Janeiro by Kohn et al. (1982). *Haemulon aurolineatum* Cuvier, 1829 is cited as a host for *A. pyriformis* (Fernandes et al., 1985) and Luque et al. (2003) recorded, for the first time, *P. brasiliensis* infected by the trematode (P=8.6% and IMI=3.4), both in the state of Rio de Janeiro. Pereira Jr. et al. (2000) reported parasitism in *M. furnieri* in Rio Grande do Sul. The specimens in this study are smaller than those collected by Kohn et al. (1982), Pereira Jr. et al. (2000) and Amato (1983b). Ribeiro et al. (2002) also reported *Brachadena pyriformis* (= *A. pyriformis*) on *P. brasiliensis* in the state of Rio de Janeiro.

The specific identification of *Lecithochirium* sp. was hampered by the specimen's sexual immaturity. Species of *Lecithochirium* are widely recorded as parasites of teleosts. Kohn et al. (2007) in their review of trematode parasites of South American fish lists 25 fish species as hosts of *Lecithochirium* spp. on the east coast of South America. *Lecithochirium microstomum* is the trematode that parasitizes the largest number of fish species in this area and is the only species of the genus known to parasitize *P. brasiliensis* in South America. The only two records for the host studied here were made by Ribeiro et al. (2002) and Luque et al. (2003) in the state of Rio de Janeiro, who reported *Lecithochirium microstomum* parasitizing the stomach of *P. brasiliensis* with P=1% and IMI=1, parasitological indices equal to those of the present study. Besides *L. microstomum*, other species of the genus, such as: *L. floridense* (Manter, 1934) Crowcroft, 1945; *L. genypteri* Manter, 1954; *L. imocavus* (Looss, 1907) Skrjabin & Guschanskaya, 1955; *L. manteri* Freitas & Gomes, 1971; *L. monticellii* (Linton, 1898) Crowcroft, 1946; *L. perfidum* Gomes, Fàbio & Rolas, 1972; *L. texanum* (Chandler, 1941) Manter, 1947 and *L. zeloticus* (Travassos, Freitas & Bührnheim, 1966) Nasir & Diaz, 1971 are also present in the study area (Kohn et al., 2007).

The differences in trematode size in relation to previous descriptions may be associated with infrapopulation variability, since the previous records on *P. brasiliensis* were made on hosts from the northern population of this species. On the other hand, the records made in the same region as this study refer to other host species, which could also result in morphometric variation.

Considering the characterization of trematode assemblages and their levels of infection as a natural tool for understanding the biology of their hosts, the trematode assemblage found in *P. brasiliensis* in southern Brazil, including three new trematode records, corroborates the existence of two populations of this host as suggested by Vargas (1976) and Paiva Filho and Zani-Teixeira (1980). There is a marked distinction between these assemblages and their respective parasitological indices between *P. brasiliensis* caught in Rio de Janeiro (northern population) and Rio Grande do Sul (southern population). The species richness of the southern population is higher (seven species) than that of the northern population (four species), but their parasitological indices are generally lower. These differences may help to mark the two distributions presented by *P. brasiliensis* in future studies in the South Atlantic.

REFERENCES

Amato J.F.R., Boeger W.A.E, Amato S.B. 1991: [Laboratory protocols: collection and processing of fish parasites]. Rio de Janeiro: University Press, Federal University of Rio de Janeiro, 81 pp.

Amato J.F.R. 1983a: Digenetic trematodes of percoid fishes of Florianópolis, southern Brazil - Homalometridae, Lepocreadiidae, and Opecoelidae, with the description of seven new species. Rev Brasil Biol. 43: 73-98.

Amato J.F.R. 1983b: Digenetic trematodes of percoid fishes of Florianópolis, shouthern Brazil - Pleorchiidae, Didymozoidae, and Hemiuridae, with the description of three new species. Rev Brasil Biol. 43: 99-124.

Braga F.M.S. 1990: [Study of *Paralonchurus brasiliensis* (Teleostei, Sciaenidae) mortality in shrimp-seven beards (*Xiphopenaeus kroyeri)* fishery area]. Bol Inst Pesca. 17: 27-35. (In Portuguese.)

Bray R.A., Gibson D.I. 1986: The Zoogonidae (Digenea) of fishes from the north-east Atlantic. Bull Br Mus Nat Hist Zool. 51: 127-206.

Bush A.O., Fernàndez J.C., Esch G.W., Seed R. 2001: Parasitism: the diversity and ecology of animal parasites. New York: Cambridge University Press, 566 pp.

Bush A.O., Lafferty K.D., Lotz J.M., Shostak A.W. 1997: Parasitology meets ecology on its own terms: Margolis et al. revisited. J Parasitol. 83(4): 575-583.

Fernandes B.M.M., Goulart M.B. 1992: First report of the genera *Macvicaria* Gibson & Bray*, 1982, Pachycreadium* Manter, 1954 and *Saturnius* Manter, 1969 (Trematoda: Digenea) in Brazilian marine fishes. Mem Inst Oswaldo Cruz. 87: 101104.

Fernandes B.M.M., Kohn A., Pinto R.M. 1985: Aspidogastrid and digenetic trematodes parasites of marine fishes of the coast of Rio de Janeiro State, Brazil. Rev Brasil Biol. 45: 109-116.

Gibson D.I., Jones A., Bray R.A. 2002: Keys to the Trematoda, Vol I. Wallingford, UK: CABI Publishing and London: The Natural History Museum, xvi + 521 pp.

Gilardoni C., Etchegoin J., Diaz J.I., Ituarte C., Cremonte F. 2011: A survey of larval digeneans in the commonest intertidal snails from Northern Patagonian coast, Argentina. Acta Parasitol. 56: 163-179.

Kohn A., Fernandes B.M.M., Cohen S.C. 2007: South American trematodes parasites of fishes. Rio de Janeiro: Editora Imprinta, 318 pp.

Kohn A., Macedo B., Fernandes B.M.M. 1982: About some trematodes parasites of *Haemulon sciurus* (Shaw, 1803). Mem Inst Oswaldo Cruz. 77: 153-157.

Luque J.L., Alves D.R., Ribeiro R.S. 2003: Community ecology of the metazoan parasites of banded croaker, *Paralonchurus brasiliensis* (Osteichthyes, Sciaenidae), from the coastal zone of the State of Rio de Janeiro, Brazil. Acta Sci, Biol Sci. 25: 273-278.

Martorelli S.R., Montes M., Marcotegui P., Alda P. 2013: Primer registro de *Diphterostomum brusinae* (Digenea, Zoogonidae) parasitando a la corvina *Micropogonias furnieri* con datos sobre su ciclo biológico. Rev Arg Parasitol. 2: 2227.

Menezes, N.A., Figueiredo, J.L. 1980: [Handbook of marine fishes from Southeastern Brazil. IV Teleostei (3)]. Sao Paulo: Museum of Zoology, University of Sao Paulo, 98 pp.

Paiva Filho A.M., Zani-Teixeira M.L. 1980: [Study of the spatial overlap of populations *Paralonchurus brasiliensis* (Steindachner, 1875) in south-east coast of Brazil between latitudes 22 ° 10'S and 29 ° 21'S (Osteichthyes, Sciaenidae)]. Rev

Bras Biol 40: 143-148. (In Portuguese.)

Pereira Jr J., Fernandes B.M.M., Robaldo R.B. 2000: Digenea (Trematoda) of *Micropogonias furnieri* (Desmarest) (Perciformes, Sciaenidae) from Rio Grande do Sul, Brazil. Rev Bras Zool. 17: 681-686.

Ribeiro R.S., Luque J.L., Alves D.R. 2002: [Quantitative aspects of parasites of banded croaker, *Paralonchurus brasiliensis* (Osteichthyes: Sciaenidae), from coast of Rio de

Janeiro state, Brazil]. Rev Univers Rural, Sér Ciênc Da Vida. 22: 151-154. (In Portuguese.)

Suchanek T.H. 1994: Temperate coastal marine communities: biodiversity and threats. Am Zool. 34: 100-114.

Vargas C.P. 1976: [Study on geographical differentiation of *Paralonchurus brasiliensis* (Steindachner, 1875) between latitudes 23° 30'S and 33°]. Msc. Dissertation, Instituto Oceanogràfico, Universidade de Sao Paulo. (In Portuguese.)

Printed by Books on Demand GmbH, Norderstedt / Germany